Solar Photovoltaic DC Calculations for Residential, Commercial and Utility Systems

Edition 1

Cover Picture: The "Two-Sun" effect may contribute to string fuses blowing on solar photovoltaic power systems.

Contents

1. Introduction

Solar photovoltaic technology has now become mainstream and is widely used around the world. All of the items needed to construct a reliable system can now be purchased from many different vendors and standards now exist in the industry to ensure that these different products can be seamlessly integrated together. The future of solar photovoltaics is bright and rapid adoption of the technology is underway.

This book is a discussion about the influences on solar photovoltaic systems and how to allow for these in the DC system design in order to build reliable systems for those who are involved in the solar photovoltaic field.

The 2008 National Electric Code states that solar photovoltaic fusing should be at least 125% x 125% of the solar module short circuit current. This gives at least 156%. But what does "at least" mean? How much higher than this figure should you go? They don't say.

This is the purpose of this book. We will look into the effects on the short circuit current value for the solar photovoltaic module and we will see approximately how much higher to go to find reliability. There's nothing worse than a solar photovoltaic power system that overheats fuses or blows fuses.

Solar photovoltaics is very different from conventional electrical system theory and because of this it has its own electrical codes that are contained in National Electric Code (NEC) Section 690 Solar Photovoltaic Systems in the USA. This book is to be used in conjunction with the National Electric Code (NEC) for residential and commercial installations and also the National Electric Safety Code (NESC) for utility installations. For construction of systems it should also be used with the local building code books.

Now lets see how much higher "at least "really means!

2. The Basics of Solar Photovoltaics

Solar photovoltaics comes in many different types:

- Monocrystalline
- Polycrystalline
- Thin film
- Other technologies

Although the technology is different between each type, they all do the same thing. All direct current (DC) solar modules generate DC electricity when exposed to sunlight. The listing above is in order of how efficiently each type will convert sunlight into electricity.

A typical direct current (DC) module will have the following electrical ratings on its label:

- Temperature adjustments
- DC Open circuit voltage (Voc)
- DC Maximum power point voltage (Vmpp)
- DC Short circuit current (Isc)
- DC Maximum power point current (Impp)
- DC Rated system voltage

All of these values are given for Standard Test Conditions (STC). Lets look at what each one of these mean:

Standard Test Conditions (STC)

Standard Test Conditions (STC) is how the solar module performs at a temperature of 25 °C, an irradiance of 1000 W/m² with an air mass 1.5 (AM1.5) spectrum. This is a standard test for all solar modules that are manufactured for the USA market that was developed by the photovoltaic industry and the government. It represents an average set of conditions that can be expected at the mid point between North and South of the contiguous forty-eight states during spring and fall with the sun perpendicular to the solar module. San Francisco, California and Wichita, Kansas are near this midpoint of 37 degrees latitude. In Asia Seoul, Korea, is near and in Europe both Sevilla, Spain and Cantania, Italy are near to 37 degrees latitude.

It is important to note that a solar module output will be continuously variable during the year and even during the day. During the changing seasons it may output less power than its rating and sometimes it may output more power. These electrical ratings are for guidance only and it is where many new photovoltaic designers make mistakes in thinking that the module will never output more power than its rating. It is important that you understand that these solar photovoltaic modules can output far more power than their label states. It can be over fifty percent more and this will need to factored into the system design.

Temperature Adjustments

Solar modules are affected by temperature, both hot and cold, and adjustments to the module ratings needs to be made for the operating temperature outside of 25 degrees Celsius. It is important when designing a system that the historical temperature minimum and maximum values are known for the area where the system is to be installed and these adjustments are factored into the design.

Open Circuit Voltage (Voc)

The open circuit voltage rating is how much voltage the module will put out with no load attached. This is an important value in order to design a system. This is the voltage to use when selecting your components and it must be adjusted for the historical minimum and maximum temperatures for the area. If more than one module is connected in series then multiply this temperature adjusted voltage by the number of modules in series to get the total maximum DC voltage of the system.

DC Maximum Power Point

The DC maximum power point is a simple concept. Power is a function of both voltage and current. The maximum power point is obtained when the current and voltage from the module when multiplied together give the maximum power figure. These values will change constantly during the day with the weather conditions. Voltage will remain relatively constant, but current will vary a lot with irradiance. The DC to AC inverter system constantly monitors the power from the solar photovoltaic DC system and automatically keeps the inverter system working at the maximum power value for the given conditions.

DC Maximum Power Point Voltage (Vmpp)

The DC maximum power point voltage (Vmpp) is the operating voltage of the solar module under load. Again this value will change with temperature and irradiance, but should only vary by about twenty percent of the STC rating during the day time.

DC Short Circuit Current (Isc)

The DC short circuit current value is the maximum current that the module will output at standard test conditions if the positive and negative terminals were connected (shorted) together. It is important to note that this value will vary a lot dependent on weather conditions and can be over fifty percent larger.

DC Maximum Power Point Current (Impp)

The DC maximum power point current is the amperage that the solar module will output at standard test conditions in normal operation. It is important to note that this value will vary a lot dependent on weather conditions and can be over fifty percent larger.

DC Rated System Voltage

This is a very important design value. It is a rating of how many modules can be safely connected together in series, this is called a solar photovoltaic module string. This system voltage should never be exceeded when adjusting for the minimum and maximum temperatures of the area that the system is being installed. This value basically limits the number of modules that can be connected in series in the system.

3. Photovoltaics and Weather

The performance of any solar photovoltaic system is dependent on the weather. The main factors that affect the system performance are clouds, irradiance, temperature, shade, latitude and how dirty the solar modules are. Let's now explore the effects of the weather in more detail:

Irradiance

Irradiance is a measure of how much sunlight the solar module is receiving. It is given in watts per meter squared or W/m². Standard Test Conditions (STC) uses a value of 1,000W/m². This value can range from 0W/m² at night through to over 1,500W/m² during a day interspersed with large fluffy clouds. This value of 1,500W/m² is larger than what you would receive in space. The reason why we can get greater values at ground level is due to what is known as the "cloud effect". Normally the sunlight is traveling in a straight line from the sun to our solar module with some atmospheric scattering. However, when clouds are present they can also reflect and can act like lenses to send some extra sunlight onto the solar modules. This extra light is converted into extra energy and this is seen largely as an increase in power from the system. This effect can be a few minutes long in duration when it occurs. The diagram on the next page demonstrates the "cloud effect".

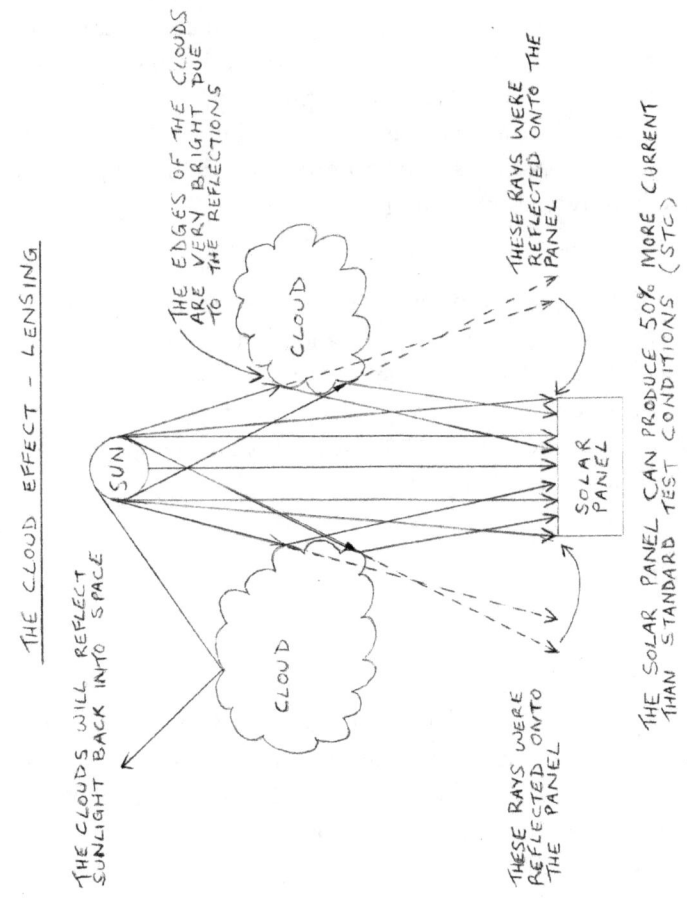

THE CLOUD EFFECT - LENSING

THE EDGES OF THE CLOUDS ARE VERY BRIGHT DUE TO THE REFLECTIONS

THESE RAYS WERE REFLECTED ONTO THE PANEL

THE CLOUDS WILL REFLECT SUNLIGHT BACK INTO SPACE

THESE RAYS WERE REFLECTED ONTO THE PANEL

THE SOLAR PANEL CAN PRODUCE 50% MORE CURRENT THAN STANDARD TEST CONDITIONS (STC)

SUN

CLOUD

CLOUD

SOLAR PANEL

Other effects on irradiance are the snow effect, the water effect (lake/ocean/wet surfaces after rain), the building effect and albedo. Snow cover, water, glass covered buildings, reflective painted buildings and roofs, and the albedo of the area surrounding the solar photovoltaic modules can reflect extra sunlight onto the solar power system. If you are installing a system in an area that has any of these, it is important to account for it. Each effect can produce an increase in power output. If you find yourself having to wear sunglasses in the solar photovoltaic system location for your eyes to be comfortable, then you probably have light reflections taking place.

During the seasons the system may operate at below the standard test conditions values and at other times it may exceed these values. During the design phase of the system you will need to assess where the greatest power need is and perhaps increase the size of the system accordingly for the particular time of year.

Air Mass

Air mass is a measurement of the the amount of atmosphere that the sunlight has to pass through to get to the ground. It varies with the seasons and also the location on the earth. Within the tropics, air mass will reach its maximum power value of 1 during summertime. Air mass 1 corresponds to the sun being directly overhead, air mass increases as the sun moves from directly overhead down to the horizon.

All USA solar modules are rated for air mass 1.5 which corresponds to a central USA location in Spring and Fall. When in a southerly location you will approach air mass 1 which will increase power output by about 13% from STC in the USA.

Locations that are at or near air mass 1 in Summer time in the USA are all Hawaiian islands, Florida and Texas. Approximately half of the continental USA is located between air mass 1 and air mass 1.5 in Spring and Fall. If you are working on systems that are located in these Southern USA states, you will get more power out of these systems due to a decreased air mass. In summertime the air mass will move closer to 1 in the continental USA.

Clouds

Clouds come in many forms. An important question is how do clouds affect irradiance on solar power photovoltaic power systems? The list below will help with understanding the effects of clouds on irradiance at air mass 1 (within the tropics in summer time):

- Clear, sunny skies will give approximately 1,130W/m^2. The transmission characteristics of the atmosphere will vary in clear skies, sometimes being relatively transparent and other times being more opaque and this affects irradiance values. Air quality is a major factor for the transmission of sunlight through the atmosphere. Particulate matter in the atmosphere will reduce the transmission level.

- Thin cirrus will give approximately 1,000W/m^2. Thin cirrus will give even and relatively stable irradiance levels due to scattering of the light.

- Thick cirrus will give approximately 750W/m^2.

- Thin clouds will give about 500W/m².

- Thick clouds will give about 250W/m². No shadows on the ground will be present

- Thick clouds with a visibly dark sky will give about 100W/m². No shadows on the ground will be present. You will not be able to see the location of the sun in the sky.

- Tall and dense broken clouds will give surges of about 1,500W/m² and reductions to about 100W/m² of irradiance due to the cloud effect. The rate and length of time for these surges and reductions is dependent on the speed of the clouds passing in front of the sun.

Temperature

Temperature will affect the system to a much lesser extent than irradiance. The cooler the system is below 25 degrees Celsius, the more power it will produce. Correspondingly, the hotter the system is above 25 degrees Celsius, the less power it will produce. Temperature can affect solar photovoltaic systems power output by about twenty percent.

Shade

It is undesirable to shade solar photovoltaic modules as it can significantly affect the performance of the system. When studying the location of where to install a system, always factor in the surroundings for shading effects. Avoid shading with solar photovoltaic power systems.

Wind

Wind will provide cooling to the photovoltaic modules and it is an aid to power production. A breezy location will provide improved performance from the system. When mounting solar modules onto racking, it is good to allow spaces between the solar modules in order to aid with cooling airflow around the modules and also to reduce wind resistance. When choosing solar modules and mounting systems, it is important to ensure that they are rated for the wind speed of the area that you are installing them into.

Altitude

A higher altitude location will improve the amount of irradiance that the system will receive, due to less scattering and absorption of the sunlight by the atmosphere. It also acts as a natural cooler of the system which further improves system performance. Generally a high altitude location will have a higher percentage of clearer skies during a year which will give a higher energy yield from the system.

Snow and Ice

Snow and ice may affect a solar module if it is faulty, causing the glass to break. It may obscure its view of the sun. Tracking systems can be affected by this and in some snowy locations it is advisable to park the tracking system facing South during these periods. The reflection from the snow will increase the power from the system in winter time.

Hail

Hail can break solar modules, so it is important to know type of hail that your area can receive. If you get large golf ball size hail, you may not want to install glass solar modules. Solar modules are tested for hail and pass the tests even if the glass module breaks. The test just ensures that the modules remains intact when broken. Glass solar modules are hard to break and normal sized hail should have no effect.

Dirt

Clean solar modules are the desirable configuration for a system. However, dust and dirt will get onto the surface of the modules and will degrade performance by up to 10% on average. Cleaning the modules is very much a function of the location where they are installed and also how dirty they are. Most people will clean on an as needed basis, generally when they are visually very dirty. Always follow the manufacturers instructions for cleaning your particular modules and remember that solar modules are operating with electricity flowing in them when exposed to light. Night time cleaning is recommended for safety.

Lightning

Lightning can affect solar modules, especially on large systems that cover fields. Good equipment grounding is the way to deal with this threat. A low resistance ground will generally dissipate lightning away from a solar module that is struck by lightning. Generally, the damage should be limited to only the solar module that was struck. If a cable is struck, then lightning surge arrestors can limit the damage in the system. These are generally installed in the inverter and on larger systems, in combiner and re-combiner boxes. Lightning may blow the string fuse/circuit breaker(s) for the module(s) struck. Install lightning protection as recommended by the manufacturers of the products used in the installation.

Seasons

We have four distinct seasons of Winter, Spring, Summer and Fall. We can word this another way as Winter Solstice (December 21), Spring Equinox (March 20), Summer Solstice (June 21) and Autumn Equinox (September 22). What does this mean to a solar power system?

- The length of the day
- The angle of the sun (air mass)
- Heating and cooling
- Rain
- Albedo

Winter solstice is the shortest day of the year and summer solstice is the longest day of the year. Spring and fall equinoxes are when day time is the same length of time as night time.

Regarding the angle of the sun in the USA, Winter Solstice is when the sun is at the lowest in the sky, or 23.5 degrees below the equator and Summer Solstice is when it is 23.5

degrees above the equator. Spring and Fall equinoxes are when the sun is directly overhead at solar noon at the equator.

For our solar power system, this means that we will produce our largest voltage in wintertime when it is the coldest and we will produce our largest current when it is summertime with peak irradiance.

The changing seasons will affect rainfall and in dry seasons you may want to schedule cleaning to keep the modules in good performance. Rain generally helps to keep the module clean naturally. Rain will also cause the albedo to change around the solar photovoltaic system and you will need to take this into account.

The albedo of the site will change during the seasons and you will need to factor this into your design. A barren snow covered field will be a lot different to one filled with corn or flowers.

There are a number of things to consider with the seasons:

- Spring & Autumn
  - The system will be operating close to standard test conditions (STC) and measured values should be close to that on the solar module label.
  - This is the most favorable time for outdoor working.
- Summer time

- The system will be hot and the DC voltage will be lower than normal.

- Ambient temperatures will be high.

- Heat and dehydration may be a problem for working on the system.

- Wintertime

  - The system will be cold and the DC voltage will be higher than normal

  - It may be too cold to work on the system

  - Frost, ice and snow may be an issue for performing maintenance.

  - Ambient temperatures will be low.

Due Diligence

It is important when designing, operating and maintaining a solar power generation system that you are aware of the annual climatic conditions to expect. Amongst the data that you should have is:

- Historic annual minimum temperature

- Historic annual maximum temperature

- Historic annual maximum wind speed

- Historic annual snow fall depth

- Historic annual hail size

- Historic annual peak irradiance

- Historic monthly irradiance

- Historic annual peak albedo
- Historic monthly albedo

With these values you will be able to make educated engineering decisions regarding the selection of your system.

4. Irradiance

Irradiance can cause problems with systems if these effects were not accounted for and designed into the system. Just as a reminder, irradiance and solar module current are generally proportional to each other. Increased irradiance will produce a corresponding increase in solar module current. There is no upper limit on solar module current output other than the string fuse blowing which is usually sized to be at least 156% of solar module short circuit current in the USA

Irradiance	Current (Isc STC)
2,000	200%
1,750	175%
1,500	150%
1,250	125%
1,000	100%
750	75%
500	50%
250	25%
100	10%
0	0%

5. Cloud Effect

The 2008 National Electric Code (NEC) states that solar photovoltaic fusing should be at least 125% x 125% of the solar module short circuit current. This gives at least 156%. But what does "at least" mean? How much higher than this figure should you go? They don't say.

What do the solar module manufacturers say? They say to fuse according to the National Electric Code. Hmm...I'm confused! They do give a little clue on their solar modules. A maximum fuse size, I think we just made some progress!

This is the purpose of this book. We will look into the effects on the short circuit current solar module value and we will see approximately how much higher to go to find reliability. There's nothing worse than a solar photovoltaic power system that overheats fuses or blows fuses.

Solar power systems that have not been designed to take in the environmental effects that surround them can be problematic. Unfortunately, some solar photovoltaic electrical design engineers are not trained in optical engineering and do not fully understand the optical environment that they are installing these systems into.

Solar Photovoltaic DC Calculations by Steven Magee

So here's what we know:

Fuse size is somewhere between 156% of the solar photovoltaic module short circuit current and the listed maximum fuse size of the solar photovoltaic module.

To find out the actual fuse size, we have to go back to the school of basic optical engineering:

Class 101: Understand the optical environment of your solar photovoltaic system.

We will now look into this in the next few chapters and class 101 starts right here.

One of the biggest irradiance effects on solar photovoltaics is the cloud effect. It hasn't been documented very well in the industry until recently. This effect needs to be accounted for and built into the design.

According to the US Department of Energy, this effect typically causes a surge on irradiance to about 150% of short circuit solar photovoltaic module current, or 1500 W/m^2. This effect is documented as lasting a few minutes in duration. This is not a short duration surge of a motor starting up, like in normal fusing. It cannot be ignored due to the length of time that the surge lasts.

So, allowing for the cloud effect, we should fuse at least 150% higher than the DC short circuit current. So here is our starting point, the fusing should be done to at least at 150% for short circuit solar module current.

Why do we use the DC short circuit value? Well, due to the maximum power point tracking system of the inverter our high performing strings may be operating near their short circuit values while the lower performing ones may be at their MPP values. The more strings that are connected to an inverter, the poorer the MPP tracking becomes on the inverter. The MPP value is a laboratory measured value of a single module and its performance will be different in the field.

So lets take a look at the 2008 NEC equation again:

125% is for DC circuit maximum continuous current

A further 125% is for fuse and cable derating

NEC minimum fuse size = 125% x 125%

= Isc x 156%

Hmm...something is wrong here. The cloud effect says 150% at least for circuit current. If we fuse for the cloud effect we get:

150% DC circuit maximum current for a few minutes

125% for fuse derating

Cloud effect fuse size = Isc x 187.5%

What does the solar module data sheet say about fusing? All solar modules have a maximum fuse size of about 2.5 times their short circuit current value. Why is this? Many years ago, someone in the industry wanted to remove blocking diodes from the strings to improve their efficiency. Improving efficiency is a great idea.

However, this creates the possibility of reverse currents occurring in the strings. Reverse currents occur due to voltage imbalances between the parallel connected strings. These imbalances occur for many reasons such as faulty solar modules, solar module shading, solar modules facing in different directions from each other, the rising irradiance levels at dusk and dawn, and so on.

Sending reverse currents through solar modules seems like a bad idea, right? Yes, it is, and that is why the fuse listed is a maximum value. This is an important concept to grasp in solar photovoltaics.

The solar photovoltaic scientists have conducted many tests to show that reverse currents below a certain level do no damage to solar modules. However, if they are allowed to get above a certain value, then they will damage the solar modules! This is where the maximum fuse value that is written on the solar modules comes into play.

The manufacturer has experimented with their solar modules and found that this maximum value of fusing allows their modules to function just fine with no damage when reverse currents occur. If the reverse currents get to large in the strings, then the fuses will blow. The fuse did its job of protecting the solar module.

If you were to increase the fuse size above the solar module maximum value, you will invalidate your warranty and also possibly damage the solar modules in the strings beyond repair.

NEVER exceed the manufacturers maximum solar module fuse size!

Okay, enough of history. Let's get back to our fuse size.

So we know that we should have string fuses that fall somewhere between 156% of the solar photovoltaic module short circuit current and the maximum fuse value on the label of the solar photovoltaic module. So exactly where in this range do we select our fuse value?

Fuse systems to at least 187.5% due to the cloud effect if you are in or near to the tropics, as the Southern USA is. But there is more to solar photovoltaic system fusing and we need to increase this value even further.

6. Albedo

Albedo is the Siamese twin of the cloud effect. They go everywhere together. Make sure that you are considering both effects when designing your system.

Albedo is the reflectivity of surfaces. A very well known effect in the world of astronomy. Unfortunately, not so well known in the world of solar photovoltaics.

Everything reflects light, even matt black surfaces reflect some low level of light, that's how we can see it. Everything our eyes can see is created from reflected light from surfaces. If our eyes can see these things, so can the solar modules. If you need to put on your sunglasses, think about the reflection effects that may cause your eyes to be uncomfortable. You will need to identify these effects as there are no sunglasses for solar modules. The solar module will just simply convert that light into more electrical current.

The diagram on the following page shows some the effects of albedo around solar photovoltaic systems.

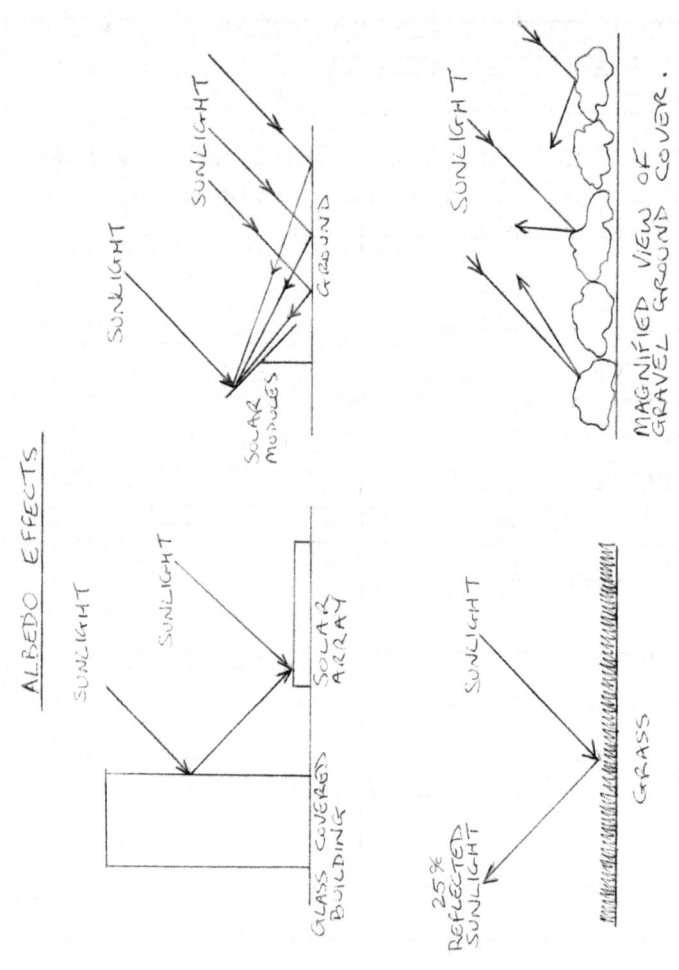

Here is a list from www.wikipedia.com that shows the albedo in various objects:

Object	Albedo
Fresh asphalt	0.04
Worn asphalt	0.12
Conifer forest	0.08 to 0.15
Deciduous trees	0.15 to 0.18
Bare soil	0.17
Green grass	0.25
Desert sand	0.4
New concrete	0.55
Ocean Ice	0.5–0.7
Fresh snow	0.80–0.90

So how do we account for this in fusing? Well, here is the fusing calculation for both the cloud effect and albedo:

Cloud effect = 150%

Albedo = ?%

Cable and fuse de-rating = 125%

Fuse size = solar module short circuit current x (cloud effect + albedo) x fuse de-rating

So we have approximations for all values except albedo and this gives:

Fuse size = Isc x (1.5 + albedo) x 1.25

So what is a reasonable value for albedo? I would suggest that albedo probably has a starting value of about 10% for solar photovoltaic systems. So this gives:

Fuse size = Isc x (1.5 + 0.1) x 1.25

= Isc x 2

So we are already fusing at double the short circuit current value of the solar module at a low level of albedo. This should be your starting point for solar photovoltaic fusing in or near to the tropics.

Albedo can get to very high levels and it is clearly important that you have a good understanding of these levels when designing a solar photovoltaic system.

Let's see how 20% albedo affects things:

Fuse size = Isc x (1.5 + 0.2) x 1.25

$= Isc \times 2.12$

And 30% albedo gives:

Fuse size $= Isc \times (1.5 + 0.3) \times 1.25$

$= Isc \times 2.25$

You will be wearing your sunglasses at this point, as your eyes will not be comfortable.

We are now approaching the 250% maximum fuse size of typical solar photovoltaic modules and we will be warming the fuse because we are operating it very close to its listed value. This will lose energy and we will have to increase the de-ratings on the fuse locations and associated equipment due to the heat dissipation.

If we were to install a system in a location with higher than 30% albedo, then we may start getting into trouble.

Highly reflective locations should be avoided in solar photovoltaics.

7. Altitude

Altitude is the best place for solar photovoltaic systems, other than Space. Altitude provides a thinner atmosphere and improves transmission of light from space to the ground. It can provide a better annual average of clear skies and it provides natural cooling of the system which improves system efficiency.

So how much more power can we expect at altitude? It all depends on the altitude and the transmission characteristics of the atmosphere. The sun generates about 1368W/m2 in space and it will be below this value. My best guess is that it is probably somewhere at about 1200 W/m2 at several thousand feet of altitude in the tropics with the sun at zenith (directly overhead).

So how would we fuse at altitude? Here is the equation below:

Altitude = 107%

Cloud effect = 150%

Albedo = 10% (or higher)

Cable and fuse de-rating = 125%

So our base fuse and cable size for an albedo of 10% would be:

Fuse size = Short circuit current x (cloud effect + albedo) x altitude x fuse de-rating

Fuse size = Isc x (1.5 + 0.1) x 1.07 x 1.25

= Isc x 2.14

We would need to adjust this figure up if we had higher values for the cloud effect or albedo.

8. High Module Current

All solar photovoltaic modules come with a tolerance rating on their power levels. It is the positive and negative adjustment values that you see listed next to the power rating of the solar module. Do not ignore the positive value.

This positive value tells us that our solar modules as supplied can actually generate more power than the labeled value. So, we need to take this into account for our fusing equation. Our equation is now:

Fuse size = Short circuit current x (cloud effect + albedo) x altitude x high performing modules x fuse de-rating

So what is a reasonable value for high performing solar photovoltaic modules? Generally +10% is listed commonly on solar module data sheets.

Using an albedo adjustment of 10% and an altitude adjustment of 107%, our fuse size is now:

Fuse size = Short circuit current x (cloud effect + albedo) x altitude x high performing modules x fuse de-rating

Fuse size = Isc x (1.5 + 0.1) x 1.07 x 1.1 x 1.25

= Isc x 2.35

We would need to adjust this higher if we had higher values for albedo or the cloud effect.

.

9. Recommended Photovoltaic Fusing

So here is the equation that you should be using to get to the correct fuse size:

Fuse size = Short circuit current x (cloud effect + albedo) x altitude x high performing modules x fuse de-rating

Cable size = 1.3 x fuse size

Smart solar photovoltaic engineers will use the maximum fuse value listed on the solar module in the USA. Even smarter solar photovoltaic engineers will recognize a solar site that has high albedo. Remember that your fuse size is also your minimum cable size too. The fuse protects the cables from overloading and possibly causing fires. You should be sizing cables to handle 1.3 times the listed fuse size due to the fuse not blowing until it exceeds 1.3 times its rated value. If you missed an effect or underestimated an effect, your cable current may end up in this range.

Fusing too low is probably the most common mistake that engineers make in the solar photovoltaic industry. It can get you into a lot of trouble.

So how do we develop a highly reflective site with a high albedo value? We need to undo history. The blocking diode is back! Or is it? I suspect that many manufacturers would not cover the solar module warranty in this situation, as if the blocking diode failed in a short circuit, then the only thing left protecting the solar photovoltaic module from reverse currents is the fuse.

So where do we go from here? There are some tricks that you can try:

- Mount your solar photovoltaic modules flat over the high albedo surface so that they cannot see the reflected light and can only see the sky.

- Screen your solar photovoltaic modules from the source of reflections

If these do not work, then you may have come to the point where you have to acknowledge that your solar photovoltaic site is unsuitable for development and walk away from it.

10. Incorrect Photovoltaic Fusing

So we will look into how our fuse works to understand incorrect solar fusing. A common fuse value in the industry is the 15 amp solar photovoltaic string fuse. Most grid connected silicon wafer solar panels use this value.

An interesting thing about a solar photovoltaic fuse is that it does not blow at the rating listed on it. It blows at a value of approximately 1.3 times higher.

So what happens if we start putting more current through the fuse? It will start to warm up and the closer it gets to the 1.3 value, the hotter it will get. At 1.3 it will be so hot that the fusable link will melt and open the circuit. Our solar photovoltaic module is now protected from potential reverse current damage.

We de-rate the fuse by 125% as fuses start to dissipate heat when you get close to their rated value. You do not want to put continuous current through fuses near their rated value otherwise your electrical losses will increase in the system. These losses are converted to heat and they will start to heat their environment.

So let's look at common industry mistakes:

Solar Photovoltaic DC Calculations by Steven Magee

Using Conventional Fuse Values

Conventional electric theory uses 125% fusing for circuits. If you fuse your system at this level, lets see what happens:

Conventional fuse size = short circuit current x 125%

The fuse will blow at approximately 1.3 times the value:

Conventional fuse opens = short circuit current x 125% x 130%

= 162.5%

So lets see what happens when some of our previously discussed effects are applied to the solar modules in the string

Summertime irradiance at or near to the tropics = 1130W/m2 or 113%

Solar module short circuit current = module short circuit current x 113%

= Isc x 113%

We have moved into the de-rating zone and the fuse will start to dissipate some energy as heat. You may have to de-rate your fuse at this point for higher enclosure ambient temperatures.

Now lets add in albedo at 10%

$$= Isc \times (1.13 + 0.1)$$

$$= Isc \times 1.23$$

The fuse is now operating at a continuous current at almost its full amperage rating. It will be getting hot.

Now lets add in a few minutes of the cloud effect at 150%. Note that the cloud effect includes the peak annual irradiance of the Sun.

$$= \text{Module short circuit current} \times (1.5 + 0.1)$$

$$= Isc \times 1.6$$

But wait, this is almost 162.5% and our fuse will be very hot. A just a little more current and the circuit will be dead and it may have overheated the equipment around it due to the heating effects of being operated in an overloaded condition. You now have a major problem and you may have to redesign and rebuild your DC system.

DO NOT FUSE AT 125% IN THE SOLAR PHOTOVOLTAIC DC CIRCUIT IN THE 48 CONTIGUOUS STATES AND HAWAII.

<u>Fusing at only 156%</u>

Now we are fusing at the 2008 National Electric Code Section 690 Solar Photovoltaic Systems value of 156% and we will apply the 1.3 factor for blowing the fuse:

Minimum NEC 690 fuse opens = short circuit current x 125% x 125% x 130%

NEC 690 fuse opens = Isc x 203%

So what happens if albedo is increased to 20% and add in high performing strings?

= module short circuit current x (cloud effect + albedo) x high performing strings

= Isc x (1.5 + 0.2) x 1.1

= Isc x 1.87

We are now operating the fuse in an overloaded condition and may be heating the associated equipment.

Lets see how 30% albedo and altitude changes things:

= module short circuit current x (cloud effect + albedo) x altitude x high performing strings

$$= Isc \times (1.5 + 0.3) \times 1.07 \times 1.1$$

$$= Isc \times 2.12$$

We have now overloaded the circuit and blown the fuse. The circuit is now dead and it may have overheated the equipment around it due to the heating effects of being operated in an overloaded condition. You now have a major problem and you may have to redesign and rebuild your DC system.

Now you can probably see why it is always best to use the manufacturers recommended maximum fuse. You should think of this as buying an insurance policy for your system reliability.

11. Photovoltaic String Theory

To understand string theory, you first have to look over the next four diagrams on the following pages. Once you have taken a look, come back and continue on:

So the first diagram is of what I would call a two internal string solar module as viewed through the glass. Basically you are looking at very large square mono-crystalline photo-diodes that have been wired internally into two series circuits.

The next diagram shows inside the junction box that is mounted to the rear of the solar photovoltaic module. Inside here you will find what is known as the "bypass diodes". These are used to bypass the internal strings if they are not producing sufficient current either from shading or internal faults.

The next diagram is the electrical diagram for the solar photovoltaic module. As you can see, it is just two series circuits of photo-diodes with a bypass diode connected across each series circuit. Each photo-diode operates at about 0.6V. Since the photo-diodes are connected in series, just like batteries, it means that each internal string on this module produces about 10 volts. Since there are two internal strings connected in series then we produce about 20 volts for this module.

Solar Module Picture

Solar Photovoltaic DC Calculations by Steven Magee

Solar Junction Box Picture

Solar Module Electrical Diagram

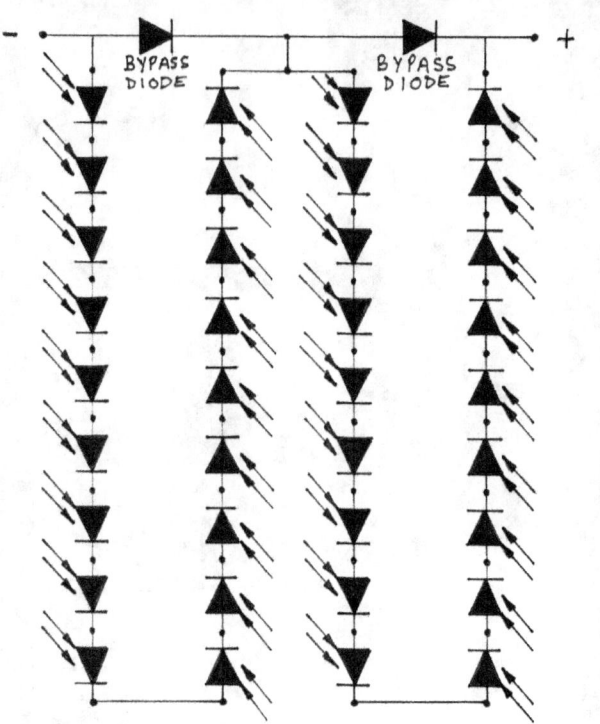

A SILICON SOLAR MODULE CONSISTS OF LARGE
PHOTO DIODE CELLS AND BYPASS DIODES

Solar Photovoltaic DC Calculations by Steven Magee

Solar Photovoltaic String Diagram

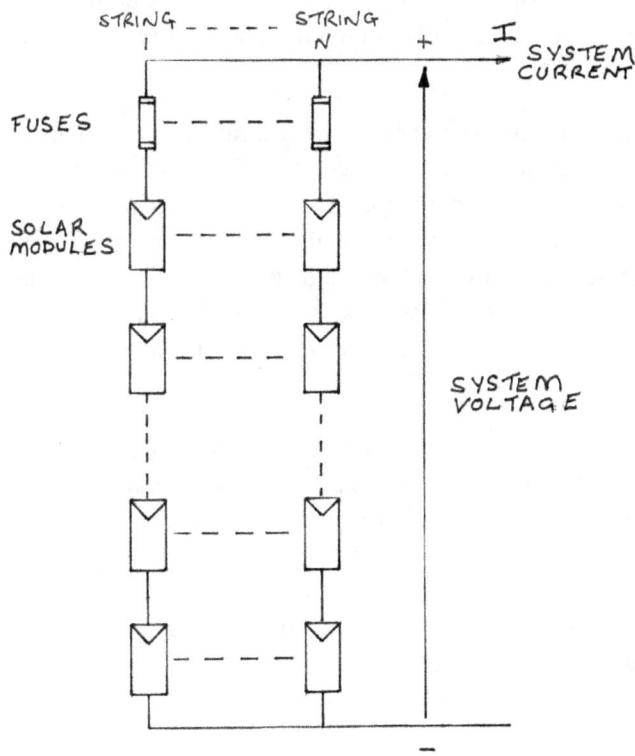

Now that we have some basic solar photovoltaic module theory, we can take a look at the next diagram. This diagram shows how we electrically wire our solar modules in a grid connected system. The first thing that we do is connect a group of solar modules in series to create a "string". You need to stay below the maximum rated voltage for the solar modules including temperature corrections for your area. So this limits the total number of modules that can be connected in series in the string.

We have put our string together and hit the voltage limit, but we need to get more power out of the system. So, to do this we start connecting solar photovoltaic strings in parallel. You keep adding equally sized strings to the system until you hit your desired system size for the inverter that it will connect on to. This creates a lot of DC current and DC voltage for your inverter system to covert into AC power.

So, is this it for string theory? Not at all, there is much more. Take a look at a typical solar module data sheet and you will see adjustments for power outside of the rated power. This can be in the rage of +10% to -10% typically. So what does this mean? Well, power is a function of voltage and current. So that +/-10% applies to both the voltage and current figures.

When you connect many solar photovoltaic modules together in a system, your are creating an averaging effect across the strings. Some strings will perform at the +10% level, others will perform at the -10% level. The rest of the strings may lie anywhere in between these values.

In an extreme case, you may have all modules either functioning at +10% or all modules functioning at -10%.

This is rare, but can happen. As such, it is important to pay close attention to the power figures for solar photovoltaic modules.

So let's get back to the normal scenario of a system of mixed power values. We know that power is a function of voltage and current. So extending this out, we have a system that has a mixed range of voltage values and current values in it. Some voltages will be higher and others will be lower. It is the same for the current.

The inverter has maximum power point (MPP) tracking built into it. This means that the inverter will keep cycling the current from the system up and down, constantly hunting for this maximum power value. It cycles the system current while watching the system voltage and when current and voltage multiplied together give the highest power, this is where it stays. This cycling is constantly occurring in the inverter.

So what does this do to our strings? Well, the ones with the highest voltage will give out more mpp current than their mpp rating, while the ones with the lower voltage are probably operating at at or below their mpp current level. Since each string is operating at a slightly different current, we need to use the maximum value of string current for our system calculations. This is the module short circuit current rating and that is why we use this value for our fuse and cable calculations.

Maximum power point tracking works best on a single solar module, and that is why AC solar modules starting to appear on the market. We are using DC solar modules and we will need a DC to AC inverter system. We could use "micro" inverters for the individual modules, but these are currently not economical on large scale systems. Installing a single inverter at each string is not currently cost effective, so we move to the system level with many parallel strings connected to the inverter system. Unfortunately, we are now operating at the lowest efficiency for maximum power point tracking.

The key with maximum power point tracking is to try and figure out where you get the best efficiency combined with the least amount of cost. Currently that is with large system inverters. This is expected to change in the future, but for now we will look into how to work with this dynamic.

So how many strings should you connect to an inverter? 10? 100? 1,000? Any amount?

Ten different engineers will give you ten different answers on this question. "How long is a piece of string?" is the same type of question. I will present my answer here based on my thoughts.

The purchase cost for the inverter per watt of conversion get less as they get larger. However, the power point tracking also degrades the larger they get and this equates to a loss in efficiency.

So at what point does this efficiency get so bad that you don't want to go any bigger. It all comes down to your worst performing strings. The worst performing strings will drag down the performance of the system.

The more strings you put in parallel, the more opportunity there is for these poor performing strings to affect the system.

I believe that the optimum value currently is to stay below one hundred strings on an inverter system. If you have these strings divided up into ten zones of ten strings at the inverter and monitor each zone, it won't take long to figure out if you have an issue with one of the zones.

By monitoring at the inverter level in zones of ten, then you do not need to install computerized string monitoring. If a string fails then you will see a 10% reduction in power from the zone when compared with the other zones. Trouble-shooting strings in zones of ten is relatively easy and quick.

So how about a 10 string inverter? I would totally recommend this route, but your large utility system will get to be a very expensive utility system! I don't recommend this due to the excessive costs of the many inverters.

So how about 1,000 string inverter? Well, you have just created a huge amount of DC power and cable capacitance in one place and you may start getting into DC interrupt current problems, as well as increasing your DC system losses due to very long DC cable runs. Troubleshooting the strings on an inverter this size would take a long time. Only the brave go down this route.

So how do high and low performing strings affect fusing?

Ahh...the mysterious string fuses that no one can understand why they keep blowing, but no faults can be detected when they are replaced. The answer is quite simple, if your system is blowing the same fuses frequently in peak combined irradiance and albedo, then it is highly likely that they are attached to the high performing strings. Mystery solved!

Well, not quite...it is also a sign that the system designer never accounted for all effects that can produce electrical current in the system and that you have now entered into a lifetime of constant high maintenance fuse replacement on your system. You will be losing energy by operating in this condition.

Fuses are very well understood electrical items with high quality controls and they do not randomly blow without reason. Imagine if your airplane had this problem!

If your system is blowing fuses, you will need to find the source of the problem and repair it. Unfortunately, you may have to rebuild your DC circuit in some extreme cases. Overheating string fuses and associated equipment is the clue for this problem.

If it is being caused by excessive albedo, then a quick and dirty fix is to put down black weed control mesh around your solar photovoltaic system. This is a lot cheaper than a rewire and may reduce your grounds maintenance bills too!

There is another source of mystery fuse blowing that can take place and that is reverse currents. This may occur in a system that has solar modules facing in different directions on the same inverter system or excessive shading issues on strings that are on the same inverter system. So keep an eye out for this if you have mystery fuse blowing, it is just the fuse doing its job of protecting the strings from internal damage from reverse currents. Generally, on this type of problem the fuses will blow either at sunrise or sunset when the voltage is rising or falling on the system which can cause voltage imbalances to occur between the strings.

Another potential source for mystery fuse blowing are the reflections that surfaces cause when the sun hits them at the right angle. Typically this will be glass, light colored painted objects nearby and water which can include ocean, lakes, swimming pools and ponds. Any surface when wet may reflect more light onto the system and this typically happens during or after rains.

So mystery string fuse blowing can be summarized as:

- Excessive current on high performing strings.
- Excessive reverse currents occurring near sunrise or sunset every day.
- Excessive reflections that may occur at only certain times during the day, month or year.

12. Inverter Sizing Calculations

The albedo effect factors into sizing the inverter system. You probably want to increase your DC Wp figure by 10% albedo as a starting point to allow for the effects of it when sizing your inverter system. You do have DC circuit losses that can offset this. It is good practice to increase the size of the inverter system so that it can convert the total system power generated by the solar modules.

If more DC power is generated than what the inverter can convert, then this power is just not used. It may have an impact on the DC system and you will need to check with your equipment suppliers about this. It will raise the maximum power point voltage of the system and will leave unused current in the solar photovoltaic modules. This excess power can be converted by increasing the inverter size or adding another inverter to the system.

So for approximate inverter sizing we get:

DC inverter size = DC Wp x DC circuit losses x (irradiance + albedo)

For example, our sample system has 10MWp DC solar photovoltaic modules, DC circuit losses of 0.85, irradiance of 1.1 and an albedo adjustment of 0.2:

DC inverter size = 10MWp x 0.85 x (1.1 + 0.2)

= 11.05 MW DC to AC inverter system.

Many people in the industry report that their systems are over performing when compared to the design predictions. This is generally a sign that the designer did not estimate the albedo effects correctly. This is common in the industry, due to a lack of awareness of albedo. Albedo is difficult to measure as it changes throughout the day, the seasons and also when it is wet.

You generally see this over performance when the irradiance is lower but the inverter system is running at full power all day long.

13. DC Interrupt Calculations

DC electrical systems have capacitance effects and in large solar photovoltaic systems, they can be very large. Let's consider what a solar photovoltaic DC circuit is comprised of:

- Lots of solar photovoltaic modules
- Lots of parallel connected solar strings
- Lots of cable
- Inverter system

Let's investigate each one of these effects in turn:

Solar Modules

Our solar modules are wired into strings. This increases our voltage to either 600 volts for residential and commercial, or 1,000 volts for utility. So our system voltage is very close to these values in wintertime

Solar Strings

Our solar strings are connected in parallel to boost the current output. So our total photovoltaic current is:

Photovoltaic short circuit current = number of parallel strings x peak solar module short circuit current

You will need to verify this with the data sheet for the solar module that you are using.

DC Wiring

There are miles of wiring in utility solar photovoltaic systems. You will need to know the capacitance of the cable wiring and also how much is used in the system to calculate the total value for capacitance.

Inverter

Inverters have very large capacitors inside them. If this can add to the short circuit fault current of the DC system, then you will need to know this value to calculate the total short circuit fault current of the DC system.

The following picture shows the electrical equivalent diagram for our DC short circuit current analysis.

DC Short Circuit Current Source Diagram

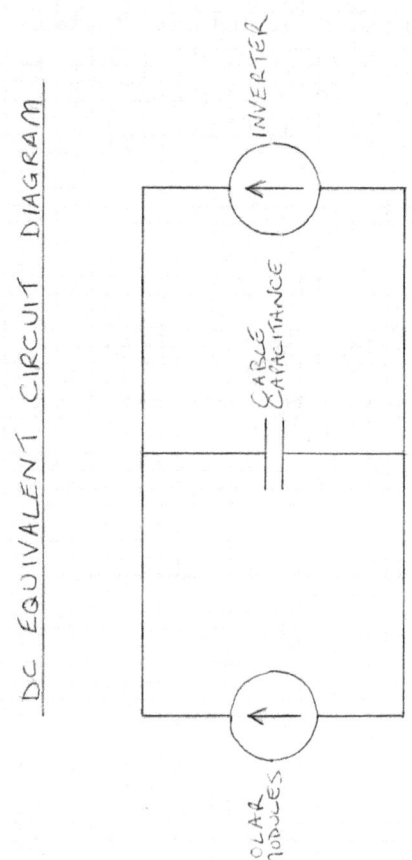

Solar Photovoltaic DC Calculations by Steven Magee

So let's see how much current a 250 kWp DC system can produce in fault conditions:

We have two extreme conditions for our system. One is increased voltage in wintertime and the other is increased current.

Here are the relevant specifications for our DC system:

- 76 strings
- 581.7V in wintertime
- 338.8V in summertime
- 250,040Wp at STC
- 8.6A solar module short circuit current
- 0.053% current temperature coefficient
- 80 degree module surface summertime temperature
- -10 degree module surface wintertime temperature
- 30% albedo
- All strings have averaged out to their labeled value.

So let us calculate our system current for summertime:

System current = ((solar module surface temp x current temperature coefficient x solar module short circuit current) + solar module short circuit current) x #strings x (cloud effect + albedo) x altitude

We will use adjustments of 1.5 for the cloud effect, 1.03 for altitude, and 0.1 for albedo:

$$= ((80 \times 0.053\% \times 8.6A) + 8.6A) \times 76 \times (1.5 + 0.1) \times 1.03$$

$$= 1,123 \text{ amps}$$

We will adjust the cloud effect value down to 1.3 to allow for a lower wintertime irradiance. And the lower wintertime value is:

$$= ((-35 \times 0.053\% \times 8.6A) + 8.6A) \times 76 \times (1.3 + 0.1) \times 1.03$$

$$= 925 \text{ amps}$$

So we have a large current available in the system from the solar photovoltaic modules. From the above you should realize that your solar photovoltaic system is changing with the seasons, another important solar photovoltaic system concept to grasp. Environmental changes that affect the peak combination of irradiance and albedo levels may increase these values on your system during the year.

The fault current is calculated for the string fuses in the combiner box. As we have seen, electrically, all strings are connected in parallel. So when we get a fault that the string fuse has to interrupt we have to analyze the total current feeding into it. We will calculate the fault current for the scenario of a short circuit on the string fuse holder. This will have almost no resistance and the entire DC system will feed into it. So how much current can feed into a short circuit with almost no resistance? Let's use a resistance value of 0.01 ohms and see the theoretical values and then we can do the actual calculation.

Using $I = V/R$ we get for maximum rated voltage of 600V:

I rated = 600V /0.01 ohm

= 60,000 A

Using the summertime voltage of 338.8V we get:

I rated = 338.8V / 0.01 ohm

= 33,880A

Using the wintertime voltage of 581.7V:

I rated = 581V / 0.01 ohm

= 58,100A

So the big question is can our system have this much current contained in it? Perhaps.

We need to find out how just exactly how big a capacitor we really built with the cabling and to do this we need to look a the cable capacitance per unit length.

Typical values for 14 AWG cable are:

Conductor DC Resistance: <2.6 ohms/1000 ft.

Nominal Capacitance: 50 pF/ft

So let's say that we have 35,000 feet of cable in our system. This gives:

= 50pF x 35,000

= 1,750,000 pF

This is insignificant and can be ignored.

How much current is available for discharge into the fault?

Let's assume that our 250 kW inverter DC maximum short circuit current is listed at 1,214A

So we get:

Short circuit fault current = photovoltaic short circuit current + inverter DC short circuit current

= 1,123A + 1,214A

= 2,337A

So there is a lot of current available in the system during fault conditions. If our DC system was ten times larger due to using a bigger inverter system, how would this look?

$$= 2{,}337A \times 10$$

$$= 23{,}370 \ A$$

A very large fault current availability. This is where you need to exercise care with very large large inverter systems.

So let us re-run the equation for an albedo of 30% and see how much more system current we get:

$$= ((80 \times 0.053\% \times 8.6A) + 8.6A) \times 76 \times (1.5 + 0.3) \times 1.03$$

$$= 1{,}263 \ amps$$

Short circuit fault current = photovoltaic short circuit current + inverter DC short circuit current

$$= 1{,}263A + 1{,}214A$$

$$= 2{,}477A$$

And at 10 times bigger for our large inverter we get:

$$= 24{,}770A$$

So how large is the interrupt current on a typical DC solar string fuse?

30,000A is typical, so as we get very large with the inverter system and solar modules, we start approaching the interrupt maximum current of the fuses. Never exceed the interrupt rating, otherwise your fuses may start to exhibit strange behaviors.

If you do exceed the interrupt rating, your fuses may suffer from violent explosions during fault conditions. You will not want to be near one of these if it happens, as it will explode like a fire cracker. If you ever see this, you have a serious problem in your circuit that you need to address before it causes a fire or someone gets injured.

Unfortunately, the most likely time for a fuse to rupture in a DC solar photovoltaic circuit is during the peak combination of irradiance and albedo which produces the peak current that will blow the fuse if the system has not been designed correctly. The fault current in the system will be very high during the fuse blowing cycle. Peak current is when the system is under the most stress and most faults are likely to occur at this time.

So you have realized that your very large DC system has exceeded the interrupt value for the fuses, what can you do to get things back into good working order? The following is a list of suggestions that may get you there:

     —   Reduce the albedo of the system

- Reduce the number of solar photovoltaic strings in the system
- Increase the interrupt rating on the fuse

So let's run one final set of calculations and for this scenario we will use the closest combiner box to the inverter system and the closest solar module to the combiner box. How large is our interrupt current if we get a short circuit on the solar module?

Looking at the cable resistance value for 14 AWG we have 2.6 ohms/1000 ft. Let's assume that we only have 10 feet of 14 AWG cable in the fault:

10' of 14 AWG resistance = 2.6 / 1000 x 10

= 0.026 ohms

We can effectively ignore the resistance for the larger cable from the combiner box to the inverter as it is so short and also that the cable is so large. So our fault current at the solar photovoltaic module becomes:

Using the summertime voltage of 338.8V for 14 AWG cable we get:

Interrupt current = 338.8V / 0.026 ohms

= 13,031A

Solar Photovoltaic DC Calculations by Steven Magee

Using the wintertime voltage of 581.7V:

Interrupt current = 581.7V / 0.026 ohms

= 22,373A

So our 30,000A interrupt rated fuse should work okay on this type of fault with the high availability of currents in the DC system.

So how about for 10 AWG cable? The resistance is listed 0.9989 per 1,000 foot of cable.

So our fault current has now changed to:

10' of 10 AWG resistance = 0.9989 ohms / 1000' x 10'

= 0.009989 ohms

We can effectively ignore the resistance for the larger cable form the combiner box to the inverter as it is so short and also that the cable is so large. So our fault current at the solar module becomes:

Using the summertime voltage of 338.8V for 10 AWG cable we get:

Interrupt current = 338.8V / 0.009989 ohms

= 33,917A

Using the wintertime voltage of 581.7V:

Interrupt current = 581.7V / 0.009989 ohms

= 58,234A

So our 30,000A interrupt rated fuse would be exceeded in both cases if there was a high availability of fault currents contained in the solar photovoltaic DC system. The fuse would most likely be damaged during the blowing cycle and could even explode. This is highly undesirable.

Be very careful with large DC inverter system designs!

14. De-rating Notes

De-rating is very, very important in the solar photovoltaics field. Outdoor equipment is operating in hot temperatures and some of it faces the sun all day long. Solar modules have current flowing in them and they dissipate heat through electrical losses, so they get even hotter than ambient temperatures.

The same is true for all electrical equipment. All electrical equipment dissipates heat when current is flowing through it. Operating this equipment in conjunction with the outdoor high ambient summertime temperatures and currents stresses it, so it is very important that you de-rate it more than sufficiently for the expected installed conditions.

De-rating is most important and must be applied to each individual component of the system for reliability:

- De-rate for highest annual ambient temperatures
- De-rate for highest expected internal enclosure temperatures
- De-rate for highest expected cable temperatures
- De-rate for highest possible currents when accounting for ALL sources of system light reflections

When in doubt, de-rate!

15. Irradiance for System Installed Area

This is the list of questions that you will need answers to so that you can accurately specify the expected maximum irradiance value for correct solar photovoltaic power system DC fuse and cable sizing:

- Maximum recorded irradiance values from cloud effects in the installation area?

- The altitude of the installed system?

- Estimated albedo of area when dry?

- Estimated albedo of area when wet?

- Seasonal changes in albedo of area?

- Any reflections from structures that may catch the sun, such as glass covered buildings?

- Any reflections from light painted roofs?

- Any reflections from snow?

- Any water reflections?

- Any rain reflections?

- Any structures that are close to the system that can reflect light onto it?

- Will there be close passing traffic?

- Did you need to wear your sunglasses on the site inspection?

- Are there any other sources of reflected light?

Once you have the answers to these questions you will be able to assess their impact on your system design and incorporate their effects on system performance to obtain a reliable solar photovoltaic power system. You will need to condense them down so that you can calculate the fuse and cable sizing:

Fuse size = short circuit current x (cloud effect + albedo) x altitude x high performing modules x fuse de-rating

Cable size = 1.3 x fuse size

As ever, if you are thinking of developing a large site, there is no substitute for a site study and a small prototype installation that produces one year of data prior to committing to the large project. An effective site study will make you successful in solar photovoltaics.

16. System Values Needed

These are the figures that you will want to have before committing to constructing a project:

- Fuse and cable sizing with a complete list of factors that were incorporated into the size selection
- Cable de-rating factors that were used throughout the system and the reasons for the selection of each de-rating figure
- Enclosure de-rating values that were used throughout the system and the reasons for the selection of each de-rating figure
- Interrupt current calculations for the circuit design for both AC and DC circuits.

With these values, you will be able to verify that the system has been designed properly. You will want to seek an independent second opinion on the values used prior to construction. Usually this would be a Professional Engineer who specializes in solar photovoltaics that would stamp the plans.

17. DC Fuse Hints and Tips

- Never fuse at 125% of the solar photovoltaic module short circuit current in the USA.

- Fusing at 156% of the solar photovoltaic module short circuit current should only ever be considered for areas that are far from the tropics.

- Fusing at the maximum fuse size as listed on the solar photovoltaic module is the safest route.

- If you want to fuse below the manufacturers listed maximum fuse value, then you will need to consider the following:

    - Local cloud effects

    - Albedo effects including reflections from wet surfaces.

    - Altitude

    - Air mass (latitude)

    - High performing solar photovoltaic modules

- Never exceed the maximum listed fuse value on the solar module otherwise you may damage your solar modules beyond repair and invalidate your warranty.

- Do not operate electrical fuses above their rating in normal operation. You should only be exceeding its rated value for a surge of short duration of several seconds or less.

- Overheating fuses, fuse holders and attached cables are a sign that your fusing is incorrect – investigate this before your system goes on fire.

Solar Photovoltaic DC Calculations by Steven Magee

- Size solar photovoltaic DC circuit cables to handle currents of 1.3 times higher than the fuse size.

- Areas with excessive albedo effects may not be suitable for solar photovoltaics and should be avoided.

- If your system suffers from excessive albedo then you can reduce it by installing black mesh weed control cloth on the ground and/or fencing around the photovoltaic modules.

- Paint objects near to the solar photovoltaic system that can reflect light onto it in darker matt paint colors.

- On reflective roofs and surfaces, mount the solar modules flat to it so that they cannot see the reflected light from the roof or surface.

- Try to eliminate pools of standing water from rainfall that are near to the system.

- Recognize sites with excessive albedo and do not develop them for solar photovoltaic systems

- Be very careful with solar photovoltaic systems in snowy climates due to reflections.

18. Common Photovoltaic Problems

- Fusing too small
- "Mystery Fuse Blowing"
- Not diagnosing "Mystery Fuse Blowing" problems correctly
- Ignoring the maximum fuse size
- Cables too small
- Overloading in voltage in wintertime
- Overloading in current during peak combined irradiance and albedo effects.
- Incorrect wind speed ratings
- Ignoring albedo effects
- Ignoring the changing seasons
- Ignoring the "Cloud Effect"
- Ignoring the "Water Effect"
- Ignoring the "Building Effect"
- Ignoring the "Snow Effect"

19. Summary

Don't let people fool you into thinking that solar photovoltaics is easy. I have heard this comment so many times in the industry. It will lure you into a false sense of security, particularly if you do not understand optical engineering.

Fuse and cable sizing is the most important aspect of solar photovoltaic system engineering and if you get it wrong, you may cause overheating effects or even worse, exploding fuses, cable insulation failures, or a fire. Your system will be unreliable.

Highly experienced electrical engineers have made mistakes in this area in the past. Now that you have read this book, you will have the knowledge to go forward and factor in all of the optical effects that can generate electrical current in the photovoltaic system and incorporate it into your design.

I would recommend any solar photovoltaic engineer to take a course in astronomy so that you can learn about the effects of air mass, albedo, altitude, light and so on that can affect solar photovoltaic power systems. These effects are very well understood in astronomy and well documented.

When designing the photovoltaic system for the contiguous USA and Hawaii, fuse at or very close to the maximum fuse value listed on the solar module. And never, ever exceed this value. If your system is repeatedly blowing the recommended maximum fuse size, then you have a problem that you will need to diagnose and fix.

Remember that your cable size has to be able to handle 1.3 times the current due to solar photovoltaic fuses not blowing until they exceed this value. Cable de-rating is very important due to this. Your solar power system will continue to convert light into current even if the irradiance from reflections is very high. The only thing that will stop excessive light conversion is the string fuse blowing.

I hope that this book answered the original question of:

"But what does "at least" mean? How much higher than this figure should you go?"

Here are the answers we came up with:

At least 156% in the USA

The recommended size to use is the listed solar photovoltaic solar module maximum fuse size

If you choose not to do this (not recommended), then you will need to use the following equation to size your fuses

Fuse size = Short circuit DC current x (local cloud effect + albedo) x altitude x high performing modules x fuse de-rating

The local "Cloud Effect" is defined as the peak annual sky irradiance for the photovoltaic system installed area.

De-rate fuses and cables for installed operating temperature conditions

Never exceed the module maximum string fuse size

All fuses must exceed the interrupt rating of the DC circuit.

All cables need to be able to handle a current of 1.3 times the fuse size.

If a potential solar photovoltaic site has excessive albedo problems that cannot be effectively dealt with, walk away from it.

As ever, National Electric Code Section 690 Solar Photovoltaic Systems is what you should be designing the DC circuit to.

Have competent and experienced solar photovoltaic professional engineers verify your designs

I would recommend further reading on solar photovoltaic systems and the reference section is a good source for this, as are my other books on the subject. My book "Solar Irradiance and Insolation for Power Systems" discusses the effects of irradiance and albedo further. This book is not a substitute for a solar photovoltaic training course and this would be a great next step.

20. Solar Photovoltaic Lexicon

- Building Effect - Peak reflected irradiance from surrounding structures.

- Cloud Effect - Peak annual sky irradiance

- DAT - Dual Axis Tracker

- EPC - Engineer, Procure and Construct

- Impp - Current at maximum power point operation

- Insolation - Time based measurement of solar irradiance. Units are watts per square meter per day

- Irradiance - Solar radiation power level. Units are watts per square meter

- Isc - DC current at short circuit operation

- kW - Kilowatt (1,000 watts)

- kWh - Kilowatt Hour (1,000 watt hours)

- m^2 - Square Meter

- MPP - Maximum power point of the optimum DC current and voltage values to produce peak power.

- MPPT - Maximum power point tracking, the inverter does this automatically to keep the DC system producing peak power.

- MW - Megawatt (1,000,000 Watts)

- MWh - Megagwatt Hour (1,000,000 watt hours)

- Net Zero - The solar photovoltaic system generates the same annual energy as is consumed annually by the residential or commercial premises where it is installed.

- PPA - Power purchase agreement

Solar Photovoltaic DC Calculations by Steven Magee

- SAT - Single axis tracker
- SLA - Site licensing agreement
- Snow Effect - Peak reflective irradiance from snow covered surfaces.
- STC - Standard test conditions
- Two-Sun Effect - When a reflection causes another sun to be seen at the same time as the real one.
- UL - Underwriters Laboratory
- UL1703 - Standard for solar module testing
- UL1741 - Standard for inverter testing
- UL4703 - Standard for photovoltaic cable
- Vmpp - Voltage at maximum power point operation
- Voc - Voltage at open circuit operation
- Water Effect - Peak reflected irradiance of wet surfaces
- Watt - Unit measure of electrical power
- W/m^2 - Watts per square meter
- $W/m^2/Day$ - Watts per square meter per day
- Wp - DC solar power at Standard Test Conditions

21. References

- NFPA National Electrical Code (NEC)

- IEEE National Electric Safety Code (NESC)

- International Building Code (IBC)

- Uniform Building Code (UBC)

- Occupational Safety and Health Administration www.osha.gov

- National Renewable Energy Laboratories www.nrel.gov

- Sandia National Laboratory http://www.sandia.gov/

- United States Department of Energy www.energy.gov

- North American Electric Reliability Corporation (NERC) http://www.nerc.com/

- Federal Energy Regulatory Commission http://www.ferc.gov/

- Solar Photovoltaics for Consumers, Utilities and Investors by Steven Magee

- Solar Photovoltaic Training for Residential, Commercial and Utility Systems by Steven Magee

- Solar Photovoltaic Design for Residential, Commercial and Utility Systems by Steven Magee

- Solar Photovoltaic Operation and Maintenance for Residential, Commercial and Utility Systems by Steven Magee

Solar Photovoltaic DC Calculations by Steven Magee

- Solar Photovoltaic Resource for Residential, Commercial and Utility Systems
- Solar Irradiance and Insolation for Power Systems

22. Author Contact

Steven Magee,

3618 S. Desert Lantern Road,

Tucson,

AZ 85735

USA

I hope that you found the book informative and please let me know about any questions or comments about the book.

I am a consultant on new solar photovoltaic projects, solar photovoltaic system troubleshooting, solar photovoltaic training, and solar photovoltaic investing for financial companies. Please feel free to contact me for any help or assistance in these areas.

You may find my other books useful:

- Solar Photovoltaics for Consumers, Utilities and Investors
- Solar Photovoltaic Training for Residential, Commercial and Utility Systems
- Solar Photovoltaic Design for Residential, Commercial and Utility Systems
- Solar Photovoltaic Operation and Maintenance for Residential, Commercial and Utility Systems

Solar Photovoltaic DC Calculations by Steven Magee

- Solar Photovoltaic Resource for Residential, Commercial and Utility Systems
- Solar Irradiance and Insolation for Power Systems